# SURVIVOR STORIES™

# TSUNAMI
## True Stories of Survival

Michael Sommers

rosen publishing's
rosen central®

New York

*To M, H, and A—with whom I survived the 1980 hurricane in Bequia.*

Published in 2007 by The Rosen Publishing Group, Inc.
29 East 21st Street, New York, NY 10010

**Library of Congress Cataloging-in-Publication Data**

Sommers, Michael A., 1966–
Tsunami : true stories of survival / Michael Sommers.—1st ed.
p. cm.—(Survivor stories)
Includes bibliographical references and index.
ISBN-13: 978-1-4042-1002-8
ISBN-10: 1-4042-1002-4 (library binding)
1. Tsunamis. 2. Survival skills. I. Title. II. Series.
GC221.2.S65 2007
363.3494–dc22

2006016523

*Printed in China*

**On the cover:** In the town of Galle, Sri Lanka, Kulmari and her son stand upon ruins that had been houses. This photo was taken almost three weeks after the Indian Ocean tsunami of December 26, 2004, devastated Sri Lanka and other Asian nations.

# CONTENTS

This photograph, taken on December 26, 2004, shows people noticing the first of several deadly tsunami waves that came crashing ashore at Koh Raya, an island resort in southern Thailand.

# INTRODUCTION: "THIS GREAT BLACK WALL..."

*I looked out . . . and saw this great big black wall coming in . . . The noise was terrific, the rolling . . . And then you heard the screaming. You look and people were stomping, trying to reach earth, trying to get out. Dogs swimming around. Then came the crash . . .*

–Kapua Heuer, "Tsunamis Remembered"
Center for Oral History, University of Hawaii at Manoa

In the United States, Hawaii experiences the greatest number of tsunamis. Small tsunamis hit the island every year, with larger, more damaging ones occurring around every seven years. However, the tragic event that made "tsunami" a household word was the recent Indian Ocean tsunami that devastated the coasts of various Asian countries on December 26, 2004. Set off by an underwater earthquake, a chain of massive waves moved across the open sea at 500 miles (805 kilometers) per hour. As the waves neared shore, they rose to the size of office buildings. When they struck land with their full force, landscapes were transformed, entire villages were wiped out, and hundreds of thousands of people were crushed, drowned, or swept out to sea—all in a matter of minutes.

Many people refer to tsunamis as "tidal waves." However, tsunamis have nothing to do with the oceans' tides. Instead, a tsunami is a series of huge waves unleashed by an undersea disturbance such as an earthquake, volcano, underwater landslide, or even a meteorite falling into the ocean. Like the ripples created when one throws a rock into still water, tsunami waves travel outward in all directions from the area of disturbance. Unlike regular waves with crests that rise up from the water's surface, tsunamis resemble vast walls of water that reach deep down to the bottom of the ocean. A tsunami can have a wavelength (the distance between wave crests) of up to 60 miles (97 km). This means that after the first wave hits the shore, an hour might go by before a second wave strikes.

Far out at sea, a tsunami spreads outward, rising only a few feet above the ocean surface. This is why a tsunami is rarely detected off-shore. It is also why deep-sea fishermen in the smallest of boats have often escaped destruction, only to return to shore and find their communities destroyed. However, as a tsunami races toward land, the shallow seabed forces the massive wall of water to slow down and rise up, in the words of one survivor, "on its hind legs like a monster." By the time the first terrifying wave reaches land, it can be 100 feet (30 meters) high, with enough power to strip sand from beaches and hurl 20-ton (18-metric-ton) rocks more than 650 feet (198 m) inland.

Those who can best attest to the horrors and devastation caused by tsunamis are the brave, strong, and often very lucky survivors who manage to escape death while witnessing the nightmarish scenes of their homes, possessions, and loved ones being swept away. This book recounts just a few such tales as it takes a look at some of the last few centuries' most destructive tsunamis and the testimonies of those who lived through them.

# 1
# THE MOST DEADLY WAVE EVER

On December 26, 2004, the world was stunned by the lethal tsunami set off by a massive Indian Ocean earthquake. Measuring 9.3 on the Richter scale, it was the second most powerful earthquake ever recorded. More than 275,000 people (at least 168,000 in Indonesia) were killed, making it the deadliest tsunami in recorded history. Worst hit were the coasts of Indonesia, Thailand, and Malaysia, as well as Bangladesh, India, Sri Lanka, and the Maldives, located thousands of miles away. Giant waves created damage as far as the East African coasts of Tanzania, Kenya, and Somalia.

## "MAMA, THE WATER!"

The greatest destruction from the tsunami occurred at Banda Aceh, Indonesia, which was at ground zero of the disaster. This coastal city was only 155 miles (249 km) from the earthquake's epicenter. Within minutes of the quake, thousands of unsuspecting people were crushed by a wall of seawater that rose to 60 feet (18 m).

People caught in the waves became like ingredients in a high-speed blender. They were smashed by chunks of concrete, cut by shards of metal and glass, and tangled up in trees. When the ocean retreated, whole communities had disappeared. Those who survived were the ones who had escaped soon enough, swum strongly enough, or held onto something stable that hadn't been swept away.

One such survivor was a thirty-three-year-old woman named Maisara. According to a November 2005 article in the *New York Times Magazine*, when the earthquake hit, she was at home with her husband and three daughters, ages eleven, nine, and three. Following the quake, Maisara's husband, a reporter, grabbed his notebook and camera and drove off to cover the story for his newspaper. Nine-year-old Ulfa was also curious as to what damage the earthquake had done. She went for a walk around the neighborhood. When she returned home ten minutes later, she was terrified. "Mama, Mama, the water!" she yelled. Ulfa grabbed her older sister by the hand and began to run. Her face was so full of fear that Maisara grabbed three-year-old Anis and followed her two daughters outside.

The crowded streets were full of people and cars. Maisara, who was overweight and couldn't move very quickly, soon lost sight of her two older daughters. With her loose flip-flops and the weight of Anis, she just couldn't keep up. Although she didn't look behind her to see what she was running from, she heard a strange roar that

The town and surrounding province of Banda Aceh, Indonesia, was the area worst hit by the 2004 Indian Ocean tsunami. Above, on December 29, three days after the giant waves struck, survivors search through the rubble of collapsed buildings to salvage what they can.

kept growing louder. Glancing over her mother's shoulder at the massive wall of approaching water, Anis shouted, "Mama, what is that?"

Suddenly, Maisara felt a great surge of water knock her legs out from under her. Clutching Anis, she fell backward and was carried toward a rice field. Then a gigantic wave struck, washing over them and pulling Anis from Maisara's grasp. Maisara was tossed and tumbled, until she briefly surfaced. Then another great wave swept her away. As she rode the water, a sharp object sawed into her leg. She smashed into a metal roof and then found herself trapped in an uprooted tree, swallowing dirty water as she fought to breathe.

A strong swimmer, Maisara broke free and found her way to the trunk of a palm tree. Exhausted, she heard a voice calling her name. A few feet away, one of her neighbors was fighting to stay afloat. Maisara stretched out her hand, but the distance was too great. For her family, she would have risked her life, but to the neighbor she said, "I'm sorry. I don't have the energy. Just pray." Then she watched the young woman sink into the dark waters.

Soon after, Maisara passed out. When she awoke, she was terrified to be in the midst of this flooded landscape. She was also stunned to discover all her clothes had been torn away; all she had left was her bra and her wedding ring. Gradually the water level diminished. Still full of fear, Maisara slid from the tree and waded through the shoulder-high water. Even though dry land was only 50 feet (15 m) away, she was exhausted. Piles of debris made the water seem like an impossible obstacle course. She began to cry out for help.

Luckily, a policeman was passing close by. As he approached Maisara, all he could see were her face and hair, completely covered in mud. Then he realized she was naked. "You have no clothes on," he exclaimed. "Just help me," begged Maisara.

The policeman found a soaked mattress pad to cover Maisara's chest. As he helped her walk, she fainted. Carrying her body so she wouldn't drown, he called for more help. It was then he saw the deep gash in her knee. It was incredible she'd been able to walk even a few steps. She also had a cut along her ear, where a flap of skin was hanging open. Finally, the policeman and three other men carried Maisara to safety, using a wooden door as a stretcher.

Maisara was brought to a doctor's house where she stayed on the roof along with other survivors. Her fears about her family's where-abouts were so intense that she didn't feel the pain of her wounds. Her injuries were serious, and she needed to go to a hospital in another city. On the way to the hospital, Maisara fell into a deep sleep. She dreamed that she was running behind her daughters and that she never caught up to them.

Throughout the month she spent in the hospital, Maisara continued to hope for news of her family. None came. Many people were consulting psychics to find missing relatives, but Maisara depended on newspaper and radio reports. Later, she traveled any place where missing girls were said to be alive, no matter their names or ages. Weeks later, she finally accepted the truth that her family was gone.

# "THE COOL WAVES"

After Indonesia, Sri Lanka, and India, Thailand was the country worst hit by the tsunami. Close to 6,000 people died. Almost half of them were foreign tourists, from fifty-six different nations, who were vacationing at the region's many beach resorts. Among them were more than 20,000 Swedes. (Of non-Asian nations, Sweden, a country of only nine million people, had the highest number of tsunami victims, with 543 deaths.)

Seventeen-year-old Christian Olsson and his family were among the many Scandinavians who had chosen to trade bitter winter weather for Thailand's tropical beaches. On the morning of December 26, according to a January 2005 CNN.com article, Christian was in his hotel room when his father called him to come outside and look at "the cool waves." When the Swedish teenager arrived on the beach, dozens of tourists were clustered around with cameras and camcorders, filming enormous waves that appeared to be miles away from the shore. However, within a few seconds, it became apparent that the waves were racing toward them at a furious speed. By the time they realized what was happening, it was too late.

Christian ran for all he was worth, but the first wave swept him up and hurled him headfirst into a wall. Unable to think, he saw a floating rooftop and managed to scramble onto it, where he lay in shock. Although he eventually rejoined his uncle and cousin, he never again saw his mother, father, grandmother, aunt, and another cousin. Within seconds, they had all been swallowed up by the giant wave. Like thousands of other young people on that day, Christian had suddenly become an orphan. According to UNICEF (the United Nations Children's Fund), the 2004 tsunami displaced, injured, orphaned, or otherwise affected more than one million children. Today, these young people are known as the "tsunami generation."

## TSUNAMI MIRACLES

Among the enormous number of tragedies caused by the 2004 tsunami, there were a few miracles as well. In Indonesia, Rizal Shahputrah heard terrified calls of warning only moments before the tsunami wiped out his village. With no time to flee, he was swept out to sea, where he would have drowned if he had not held on to an uprooted tree. In fact, surrounded by dead bodies, he clung to the tree for eight days, as ocean currents carried him 100 miles (161 km) away from shore. On the ninth day, he was spotted by sailors on a Japanese ship and rescued.

Even more incredible was the story of Jessy, an eighteen-year-old woman from tiny Pillopanja, the southernmost island off the coast of India. When the tsunami struck her small, isolated coastal village,

Survivor Rizal Shahputrah was found floating on a raft of tree branches off the shore of Banda Aceh, Indonesia. This photo was taken on January 3, 2005, shortly before he was rescued by a passing Japanese ship.

Jessy was thrown deep into the jungle. She later returned to find her village destroyed and empty. Almost all of the other 150 villagers, including her husband and baby, had been washed away. Only nine people had survived, but they had already been rescued and taken to a relief camp. Alone and terrified, Jessy ate coconuts and other jungle vegetation and waited, with increasing despair, to be saved. After forty-five days, one of the village's survivors left the relief camp and returned to check on his home. When he arrived, he found Jessy, standing upon the seashore, waiting to be rescued. Although she had lost a lot of weight and was swollen from mosquito bites, Jessy was in surprisingly good health.

Some of the most moving tsunami survivor stories were those

involving babies. Tiny, helpless, and delicate, many infants perished in the tsunami. Those who survived became symbols of hope. One such child was Hannes Bergman, an eighteen-month-old Swedish boy on vacation with his parents in Phuket, Thailand. When the waves hit the Thai resort, Hannes was carried away by the water and ended up in a pile of rubble. Soon after, an American couple and a Thai princess—who was staying in a building nearby— discovered him. With her many contacts, the princess was able to get a helicopter to come and fly Hannes to a hospital. When his photo was posted on the Internet, Hannes's uncle recognized him.

Two days after the Indonesian tsunami, Carl Michael Bergman of Sweden was miraculously reunited with his toddler son Hannes. At the time this photograph was taken, Bergman was still hoping to find his missing wife, Cecilia.

He contacted Hannes's father, Carl Michael, who had feared that his son—like his wife—would be missing forever. Carl Michael was deeply grateful to the princess, saying that not only had she saved his son, but she had saved his own soul.

Another hopeful story occurred on Hut Bay Island in India, where tsunami warnings had sent Namita Roy and her husband fleeing to the hills for safety. Eight months pregnant, Namita was due to give birth on January 15. However, the shock and terror caused by the tsunami caused her to go into early labor. On a jungly hilltop, surrounded by her husband and 700 other survivors, Namita gave birth to a healthy baby boy. Afraid of further waves, the family remained in the hills for days, surviving on bananas and coconut water, before seeking medical attention. When it came time to choose a name, the parents didn't hesitate: their son was called "Tsunami."

# 2

# EARLY TSUNAMI ACCOUNTS

Tsunamis have always been particularly common in the Pacific Ocean. This is due to the frequent shifting of the earth's tectonic plates (great pieces of the earth's crust that rub and bang against each other, creating quakes, tremors, and other disturbances that give birth to tsunamis) that takes place in this region. To explain these terrible waves, native peoples that lived along the Pacific coastline of Canada and the United States developed tales about tsunamis that were passed down over hundreds of years. Today, surviving fragments constitute some of the oldest records of tsunamis.

## "ROLLING UP OF GREAT WATERS"

Lacking the scientific knowledge that we have today, many Pacific Northwest tribes for centuries explained tsunamis through myths featuring supernatural creatures called Thunderbird and Whale. Thunderbird was a gigantic female bird whose flapping wings created thunder and great winds. Her enemy was the enormous Whale, who lived in the sea. From time to time, the two fought fierce battles. Thunderbird would wait for Whale to come to the surface of the sea and then attack him. The thrashing of Whale's great tail in the ocean created massive waves that were let loose on the peoples of the coast. According to legend, this was how tsunamis came into being.

On January 26, 1700, one of the largest earthquakes ever recorded created a tsunami that struck the west coast of Canada and the United States, as well as Japan. According to a description from the Pacific Northwest Seismograph Network Web site that has been passed down orally from generation to generation of tribes living in present-day British Columbia and Washington: "There was a shaking, jumping up, and trembling of the earth beneath, and a rolling up of great waters." The tides "rushed up at a fearful speed" and "big waves smashed into the beach," creating a "big flood."

Since tsunamis were frequent, native peoples often reacted to the earthquakes that preceded them. They escaped to higher ground or

The Salish tribe of British Columbia, Canada, made this tool and decorated it with carvings of the mythical Thunderbird and Whale.

tied their canoes to treetops and climbed in them for safety. Even so, tsunamis often came faster and stronger than expected. Giant waves wiped out villages and ripped the canoes from trees. Flooding tore up the land and changed the courses of rivers. After the water receded, survivors narrated tales of treetops filled with limbs from dead bodies and other unexpected objects. Canoes were often found deep in the middle of forests, miles from the coast.

# "WE SHALL ALL BE LOST!"

On the morning of November 1, 1755, a large earthquake struck Lisbon, the capital of Portugal. This port city, located 6 miles (10 km) from where the Tagus River flows into the Atlantic Ocean, was famous for its wealth and vast trading empire. The city had been decorated for festivities that would commemorate All Saints' Day, its greatest religious feast of the year. In fact, much of the population of 230,000 was crowded into the city's churches when the earthquake hit Lisbon (and rocked most of Europe), causing buildings to crumble and wide cracks to open up in the ground.

Tens of thousands of people who escaped the ruins and fires produced by the quake fled to the waterfront and took refuge in boats and on the empty docks. At great expense, an impressive quay of marble had just been built along the river. From here, earthquake survivors watched, fascinated and puzzled, as the

This nineteenth-century engraving depicts Lisbon, the capital of Portugal, in the throes of the devastating 1755 earthquake and subsequent tsunami waves that destroyed much of the city.

water disappeared, leaving a riverbed littered with old shipwrecks and lost cargo. Nobody was expecting a tsunami to follow the earthquake. Forty minutes after the quake, the first giant wave—estimated to be 60 feet (18 m) in height—swept up the Tagus from the open sea and roared through the city's harbor and downtown. The marble quay and all the people on it disappeared. All the boats in the harbor were destroyed and more than 20,000 citizens who had rushed to the waterfront for safety lost their lives.

Observing this disaster was the Reverend Charles Davy. His eyewitness account is excerpted on the Web site of Dr. George Pararas-Carayannis, a tsunami researcher. Following the quake, Davy had fled to the river with the crowd, where, all of a sudden, he heard people screaming, "The sea is coming in! We shall be all lost!" Davy turned toward the Tagus and was astonished to see the riverbed "heaving and swelling in the most unaccountable manner." A moment later, he saw in the distance a great body of water, rising "like a mountain. It came on foaming and roaring, and rushed towards the shore with such impetuosity, that we all immediately ran for our lives as fast as possible." Although many "were all swallowed up, as in a whirlpool, and nevermore appeared," some narrowly escaped. Davy was lucky to have grasped onto a large wooden beam that kept him afloat.

Magnificent Lisbon was in ruins, with 85 percent of its buildings reduced to rubble and close to 90,000 people (a third of the population) dead, leaving Portugal in mourning and Europe in shock. Giant waves hit the Atlantic Coast all the way up to Great Britain and as far as Finland. Great damage occurred in northern Africa, particularly Morocco, where 10,000 people died. Large waves even struck the coasts of Caribbean islands and the eastern United States.

Fortunately, among the survivors was a statesman, Marquis of Pombal, Portugal's secretary of state. A great organizer, Pombal set up an office in his carriage amid the ruins. From these headquarters,

An eighteenth-century portrait of Marquis of Pombal (1699–1782), the Portuguese statesman who took charge of Lisbon's reconstruction after the earthquake and tsunami of 1755.

he spearheaded the emergency rescue and reconstruction efforts. He also set to work investigating the scientific causes of the catastrophe. To understand more about tsunamis, he had questionnaires sent to every coastal region in Portugal asking for details about the great waves and the phenomena that accompanied them. He also invited some Japanese specialists to come to Lisbon. It was not surprising that Japanese scholars knew more about tsunamis than anybody else. They had been recording tsunamis since AD 684. Japan, to this day, has the highest tsunami rate of any country in the world.

# 3

# DISASTERS IN ASIA

In the last 1,315 years, Japan's coasts have been struck by some 195 tsunamis. This averages out to one tsunami every seven years. "Tsunami" itself is a Japanese expression meaning "harbor wave" (*tsu* means "harbor" and *nami* means "wave"). The term was coined by fishermen who returned home after having been out in deep seas to find their harbor and surrounding area completely devastated by giant waves.

## TSUNAMIS AND JAPAN

Although Japan has had its share of catastrophic tsunamis over the centuries, the most disastrous of all was the one that struck coastal villages and the port city of Sanriku on June 15, 1896. Just like in Lisbon, on the day of the tsunami, large crowds were gathered in the streets to celebrate a religious holiday. An earthquake, measuring approximately 7.2 on the Richter scale, suddenly shook the coast. Since the trembling was very mild, however, nobody was expecting the gigantic 80-foot (24 m) wave that appeared a half hour later and came crashing upon the shore. Entire villages were washed away, and 27,000 people drowned. Many of the residents who survived were fishermen who had been away at sea. When they returned to shore the following day, they were stunned to find their homes demolished and the streets covered with the bodies of their neighbors and loved ones.

One region where a majority of residents survived was the Nankaido coast close to Kyoto. Years earlier, in 1854, a tsunami had struck the region, killing 3,000 people. The number of deaths would have been higher had the quick-thinking Goryo Hamaguchi not been there. Hamaguchi, a wealthy community leader, lived in the village of Hiromura, where he had a farm of rice fields high up on a mountain facing the sea. A man of great knowledge, he recognized the warning signs of a coming tsunami. So, on December 24, when he saw the ocean pulling back from the land, Hamaguchi took action. Although his workers thought he had gone mad, he ordered them to help him set his rice fields on fire. In an area without firemen, villagers helped each other out in times of need. Hamaguchi knew his burning crops would lure the villagers up from the beach to the safety of the hills. His sacrifice saved countless lives and turned him into a living legend.

Following the tsunami, many villagers were afraid to reconstruct their homes along the beach. Hamaguchi studied the area and drew up plans for a dike to be built that would protect against

A portrait of Goryo Hamaguchi *(left)*. In October 2005, festival participants in Kainan, Japan, *(above)* light rice straws to commemorate Hamaguchi's heroic deed. He sacrificed his rice fields to direct villagers away from the devastating tsunami that hit the Japanese coast in 1854.

future tsunamis. With his personal fortune, he paid the villagers, men and women alike, to build the dike. The task took four years.

All Japanese people know Hamaguchi's tale. To this day, every year during the town's Tsunami Festival, residents pay homage to Hamaguchi's memory by bringing a handful of dirt to the dike and saying a prayer in his honor.

# An Island Erupts

On the morning of August 27, 1883, the island volcano of Krakatoa in Indonesia began to erupt. With a massive bang that was the loudest sound ever recorded, the entire island exploded. The biggest volcanic eruption in history, its force was later estimated to be almost thirty times greater than the explosion of the strongest atomic bomb. Observers in nearby ships watched as the island of Krakatoa was crushed: part of it collapsed and sunk into the sea, and other parts were vaporized. This catastrophic event produced a massive tsunami whose waves reached heights of more than 130 feet (40 m) and speeds of 450 miles per hour (724 km/h). The first four waves, which came over an interval of twenty-four hours, were immensely tall and fast. They smashed into the facing coasts of Java and Sumatra, destroying everything, including entire villages and jungles, and ultimately killing almost 40,000 people.

A German quarry manager was in his three-story office building perched atop a 100-foot-high (30 m high) hill. He saw a great wall of water surging toward him, but thought he was safe. The water smashed through his building, however, and the next thing he knew, he was being carried along on the crest of the gigantic wave, the jungle below him. Paralyzed with fear, he was sure he would drown or be crushed. Then suddenly, he saw an enormous crocodile being swept along beside him. He decided the crocodile was his only chance of survival and, imagining the beast was a life raft, hoisted himself onto its back. He pressed

This nineteenth-century lithograph shows the island of Krakatoa, in southwestern Indonesia, on the morning of August 27, 1883. A volcano erupted, created immense clouds of ash and debris, and eventually destroyed most of the island.

his thumbs tightly into the creature's eye sockets to keep himself steady and surfed upon it for 2 miles (3 km). When the wave finally broke upon a distant hill, both the man and the crocodile were tossed onto the jungle floor. Before the angry crocodile had a chance to react, the German ran off to safety, thankful for his lucky escape.

Giant waves were seen throughout the Indian Ocean, the Pacific Ocean, and as far away as the English Channel—almost 12,000 miles (19,312 km) away—where the water level rose by several inches. Victims' bodies were found floating in the ocean for weeks after. There are even documented reports of groups of human skeletons floating across the Indian Ocean and washing up on the east coast of Africa up to a year after the tsunami struck.

# BOATS TOSSED LIKE MATCHSTICKS

In Java and Sumatra, the sea flood traveled for miles inland. Forest floors, littered with seashells and great chunks of coral weighing several tons, resembled empty ocean beds. Like many other boats, a gunboat called the *Berouw* was stranded a mile inland. Its crew of twenty-eight was killed. To this day, the ship's ruins remain as a reminder of the tragedy. One of the few boats that did survive the great waves was the *Loudon*, a Dutch excursion ship that had anchored off the coastal town of Telok Betong. Passengers recall seeing a giant wave suddenly advancing toward them with great speed. Immediately, the terrified but

The tsunami unleashed by the 1883 Krakatoa volcano eruption swept miles inland, carrying boats with it. The *Berouw*, a Dutch steamer ship *(above)*, was one. Its framework can still be found.

resolute crew set sail headfirst toward the danger so that the wave's crest wouldn't break upon the ship. It had just enough time to meet the wave head on. The *Loudon* was lifted up as if it was a matchstick. Turning almost upside down, it leapt wildly and rode high over the crest of the wave, then down the other side. The wave continued on toward land, leaving passengers and crew to watch, awestruck and powerless. In a single sweeping motion, it washed over the entire town and its 5,000 residents. When the wave receded, there was nothing left but sterile, brown barrenness.

# 4
# MODERN-ERA CATASTROPHES

In the twentieth century, science finally caught up with tsunamis. Although great disasters still took place, people began to understand how and why tsunamis occurred. Armed with such knowledge, they were increasingly—though not always—able to predict coming catastrophes and protect themselves and their homes from damage.

## APRIL FOOLS

Located in the middle of the Pacific Ocean, the Hawaiian Islands have often been struck by tsunamis. Especially vulnerable is the city of Hilo, located on the Big Island of Hawaii. In 1946, the city and surrounding regions were devastated by a major tsunami caused by an earthquake near Alaska's Aleutian Islands. Massive waves hit the coast of Alaska and then struck Hawaii, where 165 people were killed. Many of them were curious children, who were fascinated when the sea receded and exposed brilliantly colored coral reefs and fish flopping on the bare ocean floor. They didn't know that this was a sign that a tsunami was on its way.

In April 1946, a massive tsunami struck Hilo, on the Big Island of Hawaii. It led to the creation of the Seismic Sea Wave Warning System, which today is known as the Pacific Tsunami Warning System.

The tsunami hit Hilo on April 1, and many people mistook early warnings for an elaborate April Fools' joke. One such Hilo resident was eighteen-year-old Mieko Browne. She says on the *National Geographic* Web site that she was just about to leave for school when she realized her shoes were dirty, so she stayed to polish them. Five minutes later, leaving the house, she heard someone shout, "Tsunami!" Both Mieko and her mother thought it was an April Fools' joke until they looked up and saw "a huge wall of dirty water," around 35 feet (11 m) in height. Just as the giant wave struck her home, Mieko's mother pulled her inside the house and slammed the door. Recalls Mieko, "It felt like we'd been hit by a train."

The wave lifted up the entire house and suddenly the Brownes were floating. Inside, Mieko was knee-deep in water. When she opened the door to her closet, the back wall of the house was gone. Beyond her hanging clothes were waves and dead fish. Meanwhile, through the windows, she could see neighbors floating by, some clinging desperately to branches, logs, and whatever else they could. Three times, the waves carried the house far out into Hilo Bay and back again, before sending it crashing into a factory wall. The Brownes managed to climb out of their house and into the factory. On the top floor, they helped some neighbors tear up burlap sugar bags. They tied the strips of burlap together to make a rope. Whenever people floated by, they tossed them the rope and pulled the victims to safety.

Today, Mieko admits that if it hadn't been for her dirty shoes, which caused her to stay home for a few extra minutes, she probably wouldn't be alive. In the meantime, the large number of deaths in Hawaii led to the creation, in 1948, of the Seismic Sea Wave Warning System. Subsequently known as the Pacific Tsunami Warning System, it tracks killer waves and provides early warning to coastal communities inhabiting the Hawaiian Islands and U.S. territories in the Pacific.

# SAVED BY BLACKBERRY BUSHES

Dr. Stuart Weinstein is a geophysicist at the National Weather Service Pacific Tsunami Warning Center, in Ewa Beach, Hawaii.

One of the twentieth century's most destructive tsunamis took place on May 22, 1960. It was created by the strongest earthquake ever recorded (9.5 on the Richter scale), whose center was located off the Pacific coast of Chile. Terrified by the shaking, which sent buildings toppling down around them, residents of coastal villages fled to their fishing boats and sought refuge on the sea. Unfortunately, within fifteen minutes of the quake, the first great wave—with a height of up to 75 feet (23 m)—swept toward the shore, swallowing up all the boats and killing more than 2,000 people. Survivors were those who headed for the hills or climbed into the branches of strong trees.

Fifteen-year-old Ramón Ramírez was one such survivor. When he saw the giant wave rushing toward him, he climbed into branches of a cypress tree. He remained there safely for hours, while raging waters swirled around the trunk. A couple of other lucky survivors were two sisters near the town of Queule, who were miraculously discovered in some blackberry bushes in the days following the disaster. The waves had apparently swept the girls into the dense, prickly bushes, where they remained sheltered from the swirling waters until residents searching for missing friends and family happened to find them.

This 1989 photograph shows Ramón Ramírez of Chile standing beside the tree that saved his life in 1960, when he was a teenager.

The tsunami left one million Chileans temporarily homeless. The giant waves swept many homes miles inland. After fleeing to higher ground, Filberto Henríquez of Queule watched as the town's houses floated away. Some of them, with their chimneys still smoking, looked like ships. Although many were destroyed, the house of Margarita Liempi was found completely intact—not even one glass or piece of china had been broken.

After striking Chile, the tsunami crossed the Pacific. Fifteen hours later, it hit Hawaii, and twenty-two hours later, it struck coast of Japan. In Hilo, Hawaii, a statewide tsunami alert issued at 6:45 on the evening of May 22. With his father and uncle, high school student Tom Goya went down to the family-owned businesses—a service station and a liquor store—to help move merchandise and important documents. With the tsunami expected to arrive around midnight, the police had closed off the whole downtown area by 11 PM.

Back home, Tom went to sleep, but was awakened at 1 AM by a noise that sounded like the end of the world. The immense roaring he heard was the sound of a 20-foot (6 m) wave crashing into Hilo Bay. The family spent the night in fear and without any electricity. At dawn, Tom, his father, and his uncle drove toward downtown. Hilo Bay was covered with giant piles of rubble. When they climbed on top of one of the piles, they couldn't believe their eyes: downtown Hilo no longer existed. Instead of houses and buildings, there was only an immense open space. There was no sign either of the gas station or the liquor store, or any of their contents. Later, by chance, they stumbled upon the liquor store's 2,000-pound (907 kg) safe near the ocean shore. Tom's uncle succeeded in opening the safe and then began to dry the money inside. Meanwhile, Tom discovered some liquor bottles from the store. Although they were sealed tightly shut, he was amazed to find that they were full of sand. Tom and his family were lucky to survive. Despite the early y, sixty-one people who failed to heed it were killed by the

A member of the Goya family sits near the wreckage of the Goya Brothers' service station *(foreground)* caused by the tsunami that struck Hilo, Hawaii, on May 22, 1960.

# "LIKE A BOMB"

Papua New Guinea is a small island nation in the Pacific that lies north of Australia. Along its coastline live the Warapu people. As fishermen, for centuries they have relied on the sea for their livelihoods. However, on July 17, 1998, the ocean brought tragedy and death.

The sun was just beginning to set when an earthquake struck 15 miles (24 km) off the island's northern coast. Although the quake itself wasn't that strong, it set off an underwater landslide. The result was a tsunami that devastated the coast, quickly killing 2,200 people. The first 40-foot-high (12 m high) wave hit the coast ten minutes after the quake.

The night was calm as Raymond Nimis, his wife, and his one-year-old daughter prepared for bed. Suddenly, their stilt house began to shake. The trembling was followed by a noise that sounded like a giant thunderclap. Seconds later, a 30-foot (9 m) wave crashed over their palm-and-bark cabin, smashing it to smithereens. In an instant, Nimis not only lost his home, but his wife and his daughter, who were swept away. Furthermore, his entire village of Arop completely vanished. Later from a hospital, Nimis described the impact of the wave as being like a bomb. Washed away by the wave, Nimis felt as if he were in a giant washing machine. It carried him inland, smashing him into trees and buildings as it snapped coconut trees and wiped out entire villages. When the wall of water finally began to rush back out to sea, it sucked along thousands of people with it. Raymond Nimis was lucky. Although he was pulled under the water numerous times, he was strong enough to grab hold of a passing wooden beam. He hung on, even when another piece of wood struck his chest.

Twenty minutes after the first of three waves hit, night had fallen. There was no moon, and in the utter darkness, with no buildings left standing, there was no way for families to look for their missing kin.

Through the night came the calls of the injured and dying. But there was no way to reach them.

Many victims were the elderly and young children who didn't have the strength to hold on to their parents or to fight the raging waves. They tried in vain to outrun the giant wall of water. Tragically, the majority of the Warapu children—an entire generation—were killed in two or three minutes. In fact, the tsunami struck on the last day of school vacation. Many children had been spending their final day with their families before returning to schools in bigger towns.

For survivors, there were no roads through the jungle to the beach areas and nowhere for rescuers to land a helicopter or airplane. People had to carry or drag the injured for hours along a jungle trail to the nearest helicopter landing pad in the town of Ramo. Afterward, many Warapu were afraid to return to their traditional coastal land. They lived on top of one of the most unstable parts of the earth's crust, where two tectonic plates had continually shifted against each other, and would certainly shift again. However, despite the threat of future tsunamis, many others couldn't turn their backs on their homeland and their fishing traditions. Said one resident in a documentary by Journeyman Pictures, "We are people of the paddle. Just because the sea hits you, you can't let go of the paddle."

# GLOSSARY

**barrenness** Emptiness, lacking in vegetation.

**burlap** A strong, coarse woven fabric made of plant fibers, often used to make bags.

**crest** The highest part of a wave.

**debris** Rubble, scattered remains of something that has been broken or destroyed.

**dike** A raised barrier of earth and stones built as a protection against floods.

**epicenter** The point on the earth's surface from which earthquake waves radiate, located directly above the actual quake.

**ground zero** The starting point of something, like a tsunami.

**lethal** Deadly.

**psychic** Someone believed to have special mental powers that may include allowing him or her to see into the future or communicate with the dead.

**quarry** An open pit from which stones are removed by digging, cutting, or blasting.

**quay** A wharf along the shore where boats are loaded and unloaded.

**Richter scale** A numerical scale (between 1 and 10) used for measuring the power of an earthquake based on ground movement and energy released.

**tectonic plate**  The two sublayers of the earth's crust that move and sometimes fracture, causing earthquakes, volcanoes, and tsunamis.

**tide**  The regular, periodic rise and fall of the sea surface resulting from the gravitational pull of the moon and sun on the earth.

**UNICEF (United Nations Children's Fund)**  An agency responsible for helping to create education and health programs for children and mothers in developing countries.

**vaporize**  To convert from a liquid or a solid into vapor.

**wavelength**  The distance between the crests of two waves.

# FOR MORE INFORMATION

Canadian Meteorological and
    Oceanographic Society (CMOS)
P.O. Box 3211, Station D
Ottawa, ON K1P 6H7
Canada
(613) 990-0300
Web site: http://www.cmos.ca


Federal Emergency Management
    Association (FEMA)
500 C Street SW
Washington, DC 20472
(800) 621-FEMA (621-3362)
Web site: http://www.fema.gov/hazard/tsunami/index.shtm


International Tsunami Information Centre (ITIC)
Pacific Guardian Center, Mauka Tower
737 Bishop Street, Suite 2200
Honolulu, HI 96813
(808) 532-6422
Web site: http://www.tsunamiwave.info

National Oceanic and Atmospheric Administration (NOAA)

14th Street & Constitution Avenue NW, Room 6217

Washington, DC 20230

(202) 482-6090

Web site: http://www.tsunami.noaa.gov

Pacific Tsunami Museum

P.O. Box 806

Hilo, HI 96721

(808) 935-0926

Web site: http://www.tsunami.org

Pacific Tsunami Warning Center

91-270 Fort Weaver Road

Ewa Beach, HI 96706

(808) 689-8207

Web site: http://www.prh.noaa.gov/ptwc

# WEB SITES

Due to the changing nature of Internet links, Rosen Publishing has developed an online list of Web sites related to the subject of this book. This site is updated regularly. Please use this link to access the list:

http://www.rosenlinks.com/ss/tsun

# FOR FURTHER READING

Bonar, Samantha. *Tsunamis* (Natural Disaster). Mankato, MN: Capstone Press, 2001.

Buck, Pearl S. *The Big Wave*. Rev. ed. New York, NY: Harper Trophy, 1986.

Drohan, Michele Ingber. *Tsunamis: Killer Waves* (Natural Disasters). New York, NY: PowerKids Press, 1999.

Fredericks, Anthony D. *Tsunami Man: Learning About Killer Waves with Walter Dudley*. Honolulu, HI: University of Hawaii Press, 2002.

Thompson, Luke. *Tsunamis* (Natural Disasters). New York, NY: Children's Press, 2000.

Wade, Mary Dodson. *Tsunami: Monster Waves*. Berkeley Heights, NJ: Enslow Publishers, 2002.

# BIBLIOGRAPHY

Atwater, Brian F, et al. "Surviving a Tsunami—Lessons from Chile, Hawaii, and Japan." U.S. Geological Survey, Circular 1187, Version 1.0, 1999. Retrieved March 2006 (http://pubs.usgs.gov/circ/c1187/).

BBC Science and Nature: Hot Topics. "Natural Disasters." Retrieved March 2006 (http://www.bbc.co.uk/science/hottopics/ naturaldisasters/index.shtml).

Bearak, Barry. "The Day the Sea Came." *New York Times Magazine*, November 27, 2005. Retrieved March 2006 (http://www.nytimes. com/2005/11/27/magazine/27tsunami1.html?ei=5090&en= d394c55cf2f02e7b&ex=1290747600&partner=rssuserland&emc= rss&pagewanted=all).

CNN.com. "After the Tsunami." Retrieved March 2006 (http://www. cnn.com/SPECIALS/2004/tsunami.disaster).

CNN.com. "Teen Tells of Losing 5 of Family." Retrieved March 2006 (http://www.cnn.com/2005/WORLD/europe/01/07/losing.family/ index.html).

Journeyman Pictures. "Papua New Guinea—Tsunami: The Survivor's Story." Retrieved March 2006 (http://www.journeyman.tv/ ?lid=18017).

Ludwin, Ruth. "Draft: Cascadia Megathrust Earthquakes in Pacific Northwest Indian Myths and Legends." The Pacific Northwest

Seismograph Network. December 29, 1999. Retrieved March 2006 (http://www.pnsn.org/HIST_CAT/STORIES/draft1.html).

NationalGeographic.com. "I Survived a Tsunami." Retrieved March 2006 http://www.nationalgeographic.com/ngkids/9610/kwave/survive.html).

National Geographic News. "Tsunamis: Facts About Killer Waves." Updated January 14, 2005. Retrieved April 2006 (http://news.nationalgeographic.com/news/2004/12/1228_041228_tsunami.html).

Pacific Tsunami Museum. "Student Guide." Retrieved March 2006 (http://www.tsunami.org/students.htm).

Pararas-Carayannis, George, Dr. "The Great Lisbon Earthquake and Tsunami of 1 November 1755." The Tsunami Page of Dr. George P.C. Retrieved March 2006 (http://www.drgeorgepc.com/Tsunami1755Lisbon.html).

Pendick, Daniel. "A Deadly Force." Savage Earth: Waves of Destruction: Tsunami. PBS Online. Retrieved March 2006 (http://www.pbs.org/wnet/savageearth/tsunami/index.html).

Sakurai, Nobuo. "A Tsunami Is Coming! Don't Put Out the Inamura Fire." Tokyo, Japan: Urban Disaster Research Institute, 2005. Retrieved March 2006 (http://www.tokeikyou.or.jp/bousai/english/files/Text_116_English+1page.pdf).

University of Hawaii at Manoa Center for Oral History. "Historical Events: Tsunamis Remembered." Retrieved March 2006 (http://www.oralhistory.hawaii.edu/pages/historical/tsunami.html).

# INDEX

# ABOUT THE AUTHOR

A writer and journalist, Michael Sommers has a BA in literature from McGill University and a master's in history and civilizations from the École des Hautes Études en Sciences Sociales, in Paris, France. Although he has never witnessed a tsunami, Sommers has walked on volcanoes and been caught in the full force of a Caribbean hurricane.

# PHOTO CREDITS

Designer: Tahara Anderson; Editor: Elizabeth Gavril
Photo Researcher: Amy Feinberg